BAOFENG RADIO BASICS SIMPLIFIED

Your Essential Guide for Beginners to Get Started with Two-Way Radio and Guerrilla Communication.

Jenny Holmes

All rights reserved © Jenny Holmes 2024

Table Of Contents

INTRODUCTION.. 5
CHAPTER 1: Unveiling Your Baofeng Radio..............9
   Unboxing and Setting Up: Your Baofeng at First Glance... 9
      Powering Up for Success: Battery Essentials... 11
CHAPTER 2: Demystifying Frequencies and Channels.. 16
   Understanding the Radio Spectrum: Waves, Frequencies, and Wavelengths............................. 16
   Channels: Your Highways on the Airwaves............ 18
      Privacy on the Airwaves.................................. 21
CHAPTER 3: Mastering the Art of Radio Communication... 24
   Radio Etiquette: The Unspoken Rules of Respectful Communication...24
   Receiving Messages: Active Listening on the Radio.. 28
CHAPTER 4: Unveiling the Features of Your Baofeng Radio...32
   Essential Baofeng Functionalities for Everyday Use.. 32
      Squelch Control: Eliminating Unwanted Noise. 35
      Busy Lockout: Preventing Transmission on Occupied Channels... 37
      Memory Scan for Frequently Used Frequencies.. 39
      Cross-Band Repeat: Extending Communication

Range (Concept Only)..41

CHAPTER 5: Expanding Your Baofeng's Potential with Accessories..44

Enhancing Your Communication Experience: Essential Accessories..44

Programming Cables for Customization................49

Optimizing Range with Antennas: Matching Your Needs..51

Choosing the Right Antenna: Benefits, Limitations, and Safe Use..54

CHAPTER 6: Safety and Regulations: Using Your Baofeng Responsibly..57

Understanding Licensing Requirements: Operating Legally..57

The Importance of Obtaining a License (e.g., FCC Amateur Radio)..58

Different License Types and How to Get One..60

Consequences of Operating Without a License...61

Designated Frequencies for GMRS, FRS, MURS (Depending on Region)..63

Battery Safety and Disposal: Protecting Yourself and the Environment..65

CHAPTER 7: Keeping Your Baofeng Running Smoothly: Maintenance and Troubleshooting..........70

Basic Maintenance: Caring for Your Baofeng Radio..70

Cleaning Instructions for Radio Body, Antenna, and Microphone..70

Troubleshooting Common Issues: Getting Your Radio Back on Track..73

When to Seek Help from Qualified Technicians......75
CONCLUSION.. **77**
Recap and Key Takeaways: The Power of Baofeng Radios..77
Encouragement for Continued Learning and Responsible Use..80
GLOSSARY OF TERMS: Defining Technical Lingo. 82

INTRODUCTION

In a world where instant communication is king, there's something timeless and empowering about the simplicity of two-way radios. Whether you're embarking on a wilderness adventure, coordinating a community event, or simply staying connected with friends and family, the humble two-way radio remains a reliable companion in an ever-evolving digital landscape.

Welcome to the World of Two-Way Radios. This book invites you to step into a realm where communication knows no bounds. Here, distances are bridged with the push of a button, and conversations unfold with the crackle of static and the clarity of purpose. It's a world where immediacy reigns supreme, where words traverse vast expanses of space in mere milliseconds, binding individuals together in a network of shared experiences and shared goals.

But amidst the sea of options, one name stands out: Baofeng. In "Introducing Baofeng Radios: Affordability, Features, and Fun," we delve into the heart of what makes these devices so special. More than just tools for communication, Baofeng radios represent a fusion of affordability, functionality, and sheer enjoyment. They're the Swiss army knives of the airwaves, packed with features that cater to both novice users and seasoned enthusiasts alike.

With Baofeng radios, the world becomes your playground. From their sleek and ergonomic designs to their intuitive controls and robust construction, these devices are engineered to withstand the rigors of real-world use while providing unparalleled performance. And with a price point that won't break the bank, they offer entry into the world of two-way communication without barriers or limitations.

But perhaps the most compelling aspect of Baofeng radios lies in the sense of adventure they inspire. Whether you're exploring the great outdoors, participating in emergency response efforts, or simply engaging in leisurely conversations with fellow enthusiasts, these devices open up a world of possibilities. With features like long-range communication, programmable channels, and advanced scanning capabilities, Baofeng radios empower users to take control of their communication destiny and forge connections that transcend distance and time.

In "The BAOFENG RADIO BASICS: Your Essential Guide for Beginners to Get Started with Two-Way Radio Communication," we embark on a journey of discovery, demystifying the world of two-way radios and unlocking the potential of Baofeng devices. From the fundamentals of radio operation to advanced techniques for maximizing performance, this book is your passport to a new realm of communication possibilities.

So whether you're a seasoned radio aficionado or a curious newcomer, join us as we explore the endless horizons of two-way radio communication with Baofeng radios. The adventure awaits – are you ready to answer the call?

CHAPTER 1: Unveiling Your Baofeng Radio

Unboxing and Setting Up: Your Baofeng at First Glance

Identifying Parts and Accessories

As you embark on your journey with your Baofeng radio, the first step is to familiarize yourself with its various components and accessories. Upon unboxing, you'll encounter essential items such as the radio unit itself, antenna, battery pack, belt clip, wrist strap, user manual, and possibly additional accessories depending on the model you've chosen.

Take a moment to inspect each component, noting their functions and how they fit together to form a cohesive communication tool. The antenna, for example, is crucial for maximizing signal reception and transmission efficiency, while the battery pack provides the power necessary to keep your radio operational.

Getting Powered Up: Battery Installation and Charging

Once you've identified the parts and accessories, it's time to power up your Baofeng radio. Begin by installing the battery pack according to the manufacturer's instructions, ensuring a secure fit to prevent any interruptions in power supply during operation.

Next, connect the radio to the appropriate charging source using the provided charger or USB cable. Follow the recommended charging guidelines outlined in the user manual to avoid overcharging or damaging the battery. In most cases, a full initial charge is recommended to maximize battery life and performance.

As your Baofeng radio charges, take the opportunity to familiarize yourself with its basic operation. Practice turning the device on and off, adjusting the volume levels, and navigating the keypad to access essential functions. By mastering these fundamental operations, you'll be better prepared to dive into more advanced features and capabilities later on.

Powering Up for Success: Battery Essentials

Understanding Battery Types and Capacities

The battery is the lifeblood of your Baofeng radio, providing the power necessary to keep you connected in the field. Understanding the different types of batteries available for your device is essential for optimizing performance and longevity.

Baofeng radios typically utilize rechargeable lithium-ion (Li-ion) batteries, prized for their high energy density, lightweight design, and ability to hold a charge over multiple cycles. These batteries come in various capacities, measured in milliampere-hours (mAh), which determines how long your radio can operate on a single charge.

When selecting a battery for your Baofeng radio, consider factors such as capacity, voltage, and compatibility with your specific radio model. Higher capacity batteries offer extended operating times but may add bulk and weight to your device. Conversely, lower capacity batteries may be more compact and lightweight but may require more frequent recharging.

Safe Battery Practices: Charging, Storage, and Disposal

To ensure the longevity and safety of your Baofeng radio battery, it's essential to follow best practices for charging, storage, and disposal.

When charging your battery, always use the provided charger or a reputable third-party charger designed specifically for your Baofeng radio model. Avoid using incompatible chargers or makeshift charging methods, as they can damage the battery and pose a fire hazard.

Additionally, adhere to recommended charging times and avoid overcharging your battery, as this can degrade its performance and shorten its lifespan. Once fully charged, disconnect the charger to prevent unnecessary strain on the battery.

When storing your Baofeng radio battery, keep it in a cool, dry place away from direct sunlight and extreme temperatures. Avoid storing the battery in a fully discharged state, as this can lead to irreversible capacity loss over time. If you anticipate prolonged periods of inactivity, consider partially charging the battery before storage to maintain optimal performance.

Finally, when it comes time to dispose of your Baofeng radio battery, do so responsibly in accordance with local regulations and recycling programs. Many electronics retailers and recycling centers offer battery recycling services to ensure proper disposal and minimize environmental impact.

By following these battery essentials, you can maximize the performance, longevity, and safety of your Baofeng radio, ensuring reliable communication when you need it most.

CHAPTER 2: Demystifying Frequencies and Channels

Understanding the Radio Spectrum: Waves, Frequencies, and Wavelengths

VHF vs. UHF Bands: Applications and Range Considerations

To truly grasp the intricacies of two-way radio communication, it's essential to understand the underlying principles of the radio spectrum. At its core, the radio spectrum is a vast expanse of electromagnetic waves, each characterized by its frequency and wavelength.

VHF (Very High Frequency) and UHF (Ultra High Frequency) are two primary bands within the radio spectrum, each with its unique characteristics and applications. VHF frequencies typically range from 30 MHz to 300 MHz, while UHF frequencies span from 300 MHz to 3 GHz.

In practical terms, VHF frequencies are well-suited for applications requiring long-distance communication over open terrain. Their longer wavelengths enable them to travel farther and penetrate obstacles such as buildings and vegetation with greater ease. As a result, VHF radios are commonly used in outdoor settings such as wilderness expeditions, search and rescue operations, and maritime communication.

On the other hand, UHF frequencies excel in urban environments and indoor settings where obstacles and interference are more prevalent. Their shorter wavelengths allow for better signal penetration through dense materials and urban clutter, making them ideal for applications such as public safety, construction, and manufacturing.

When selecting a radio for your specific needs, consider the advantages and limitations of each frequency band. VHF radios offer superior range in open environments but may struggle in densely populated areas with high levels of interference. Conversely, UHF radios excel in urban settings but may experience decreased range and performance in remote or rural areas.

Channels: Your Highways on the Airwaves

Pre-programmed Channels and Scannable Frequencies

Channels serve as the virtual highways on the airwaves, providing organized pathways for communication between radio users. In the world of two-way radios, channels are typically pre-programmed frequencies or frequency ranges allocated for specific purposes such as public safety, amateur radio, and commercial use.

Baofeng radios often come pre-programmed with a selection of commonly used channels, allowing users to quickly access essential frequencies without the need for manual programming. These channels may include emergency services, weather broadcasts, and common amateur radio bands, providing immediate access to vital information and communication networks.

In addition to pre-programmed channels, many Baofeng radios feature scanning capabilities, allowing users to scan through available frequencies to identify active channels or monitor for incoming transmissions. This feature is particularly useful in dynamic environments where communication needs may change rapidly, such as during emergency response operations or large-scale events.

User-Programmed Channels (for Licensed Users)

For users with specific communication requirements or operating within licensed radio services, Baofeng radios offer the flexibility to program custom channels tailored to their needs. Whether you're a licensed amateur radio operator, a member of a public safety agency, or part of a commercial organization, user-programmable channels allow you to fine-tune your radio's frequency settings for optimal performance.

By programming custom channels, licensed users can ensure compliance with regulatory requirements, avoid interference from unauthorized users, and maintain secure communication channels for sensitive operations. Baofeng radios provide intuitive programming interfaces and software tools to simplify the process of creating and managing user-programmable channels, allowing users to customize their radios to suit their specific communication needs.

Privacy on the Airwaves

CTCSS/DCS Tones: Reducing Interference on Shared Frequencies

In crowded radio environments where multiple users share the same frequencies, maintaining privacy and minimizing interference is paramount. Continuous Tone-Coded Squelch System (CTCSS) and Digital-Coded Squelch (DCS) tones offer a solution by providing a means of selectively filtering out unwanted transmissions based on pre-defined codes.

CTCSS tones, also known as sub-audible tones or privacy codes, are low-frequency audio tones transmitted along with the voice signal. Receivers equipped with CTCSS decoding capabilities will only unmute and pass audio when the transmitted tone matches the programmed tone, effectively ignoring transmissions from other users without the correct tone.

Similarly, DCS tones use digital codes instead of analog tones to achieve the same purpose, providing an additional layer of privacy and interference rejection. DCS tones offer greater reliability and resistance to noise compared to CTCSS tones, making them ideal for environments with high levels of interference or where privacy is a priority.

By enabling CTCSS or DCS tones on your Baofeng radio, you can minimize the impact of unwanted transmissions and ensure clear communication with your intended recipients. Whether you're coordinating with a team in the field or conversing with fellow enthusiasts on a shared frequency, these privacy features help maintain order and efficiency on the airwaves.

CHAPTER 3: Mastering the Art of Radio Communication

Radio Etiquette: The Unspoken Rules of Respectful Communication

Greetings, Over/Out Signals, and Avoiding Interruptions

Effective radio communication relies not only on the clarity of your message but also on adherence to established etiquette and protocol. By observing the unspoken rules of radio etiquette, you can ensure respectful and efficient communication with fellow users on the airwaves.

Begin each transmission with a courteous greeting, such as "Good morning" or "Hello," followed by the call sign or identifier of the station you're addressing.

This simple gesture sets a positive tone for the interaction and establishes rapport with the receiving party.

During the course of a conversation, use the terms "Over" and "Out" to signal the status of your transmission. "Over" indicates that you've finished speaking and are awaiting a response, while "Out" signifies the conclusion of the communication and that no further response is expected. Avoid unnecessary interruptions by waiting for the appropriate moment to transmit and allowing other users to finish their transmissions before speaking.

In situations where multiple parties are participating in a conversation, use clear and concise language to avoid confusion and ensure everyone has an opportunity to be heard. Practice active listening and refrain from transmitting unless you have something relevant to contribute to the discussion.

By following these simple guidelines, you can foster a culture of respectful communication on the airwaves and minimize disruptions for yourself and others.

Calling All Stations: Making Clear and Effective Transmissions

Call Procedures: Identifying Yourself and Your Message

When initiating a transmission on the radio, it's essential to follow proper call procedures to ensure your message is received clearly and accurately by the intended recipients.

Begin by identifying yourself and your station using your call sign or other identifier specified by regulatory authorities or organizational policies. This allows other users to know who is transmitting and helps maintain accountability and clarity in communication.

Once you've established your identity, clearly state the purpose of your transmission and provide any relevant information in a concise and organized manner. Avoid unnecessary preamble or filler words that can detract from the clarity of your message and waste valuable airtime.

When addressing specific stations or individuals, use their call signs or identifiers to ensure your message reaches the intended recipients. Be mindful of your tone and demeanor, maintaining professionalism and courtesy at all times, regardless of the nature of the communication.

Finally, conclude your transmission with a clear and definitive sign-off, such as "This is [Your Call Sign], out." This signals to other users that you've finished speaking and allows them to respond or continue the conversation as needed.

By mastering the art of clear and effective transmissions, you can streamline communication on the airwaves and ensure your messages are received and understood by those who need to hear them. Whether you're conveying critical information in an emergency situation or engaging in casual conversation with fellow enthusiasts, clear communication is the key to successful radio operation.

Receiving Messages: Active Listening on the Radio

Optimizing Reception: Volume and Minimizing Background Noise

Active listening on the radio begins with optimizing your reception to ensure clear and consistent communication. Start by adjusting the volume settings on your radio to a comfortable level that allows you to hear incoming transmissions without straining or causing discomfort.

In noisy environments or situations with high levels of background noise, consider using headphones or an external speaker to improve clarity and reduce distractions. Position the radio antenna for optimal signal reception, avoiding obstructions that may interfere with reception quality.

Minimize background noise by selecting a quiet location for radio operation whenever possible. Avoid transmitting or receiving near sources of interference such as electronic devices, power lines, or machinery, as these can introduce unwanted noise and distortions into your transmissions.

Radio Response Etiquette: Waiting Your Turn and Clear Acknowledgments

When receiving messages on the radio, practice patience and courtesy by waiting for the appropriate moment to respond. Avoid interrupting ongoing transmissions or conversations unless you have urgent or relevant information to contribute.

Listen actively to incoming transmissions, focusing on the content of the message and any instructions or requests conveyed by the sender. Take notes if necessary to ensure you capture important details and follow-up actions.

When acknowledging received messages, use clear and concise responses to confirm receipt and understanding. Avoid unnecessary chatter or acknowledgments that can clutter the airwaves and disrupt communication flow.

If multiple parties are involved in a conversation, wait for a break in transmission before interjecting with your response. Use brief pauses to indicate your intention to transmit, allowing others to finish speaking before transmitting your message.

By practicing active listening and observing proper radio response etiquette, you can ensure effective communication on the airwaves and contribute to a positive and productive radio environment. Whether you're engaged in casual conversation or coordinating critical operations, clear reception and thoughtful responses are essential for successful radio communication.

CHAPTER 4: Unveiling the Features of Your Baofeng Radio

Essential Baofeng Functionalities for Everyday Use

Dual-band/Dual-display (on Specific Models)

One of the standout features of many Baofeng radios is their dual-band and dual-display capabilities, available on specific models. This functionality allows users to monitor and transmit on two different frequency bands simultaneously, providing increased flexibility and versatility in communication.

With dual-band capability, Baofeng radios can operate on both VHF and UHF frequencies, expanding the range of available channels and communication options. This is particularly useful in environments where multiple frequency bands are in use, such as amateur radio bands, public safety channels, and commercial radio services.

Dual-display functionality further enhances the user experience by allowing simultaneous display of information from both frequency bands. This enables users to monitor activity on each band in real-time, easily switch between channels, and quickly access essential information without the need for manual adjustments.

Whether you're monitoring weather broadcasts on VHF while coordinating with a team on UHF, or participating in a cross-band repeater operation, dual-band/dual-display functionality empowers users to stay connected and informed across multiple frequency bands.

VOX (Voice Operated Transmit): Hands-Free Communication

Hands-free communication is essential in situations where manual operation of the radio may be impractical or unsafe. Baofeng radios equipped with Voice Operated Transmit (VOX) functionality offer a convenient solution, allowing users to transmit voice signals without the need to press a push-to-talk (PTT) button.

VOX operates by detecting the user's voice and automatically activating transmission when audio is detected above a predefined threshold. This enables users to communicate seamlessly while keeping their hands free for other tasks, such as operating equipment, navigating terrain, or performing emergency procedures.

By adjusting VOX sensitivity settings, users can fine-tune the activation threshold to suit their specific environment and speaking volume. This ensures reliable performance in varying conditions, from quiet indoor settings to noisy outdoor environments.

Whether you're engaged in outdoor activities such as hiking or cycling, coordinating events or operations where hands-free communication is essential, or simply looking for added convenience in everyday use, VOX functionality enhances the usability and versatility of Baofeng radios.

Squelch Control: Eliminating Unwanted Noise

In radio communication, unwanted noise and interference can detract from the clarity of received signals and disrupt communication flow. Baofeng radios feature Squelch Control functionality, allowing users to adjust the sensitivity of the receiver to filter out background noise and interference.

Squelch works by muting the audio output of the receiver when the incoming signal falls below a predefined threshold, effectively eliminating low-level noise and interference while allowing stronger signals to pass through unaffected.

By adjusting the squelch level, users can customize the radio's sensitivity to match the prevailing conditions and optimize reception quality. In noisy environments, a higher squelch setting can help suppress background noise and improve clarity, while in quieter settings, a lower squelch setting ensures maximum sensitivity to weak signals.

Squelch control is especially useful in scenarios where multiple users share the same frequency, such as in amateur radio repeater systems or crowded public safety channels. By fine-tuning the squelch settings, users can minimize distractions and focus on clear and effective communication.

In conclusion, essential Baofeng functionalities such as dual-band/dual-display, VOX, and squelch control enhance the usability, versatility, and performance of Baofeng radios in everyday use. Whether you're a casual user, outdoor enthusiast, or professional communicator, these features empower you to stay connected and informed in any situation.

Busy Lockout: Preventing Transmission on Occupied Channels

Busy Lockout is a valuable feature found in many Baofeng radios that helps prevent accidental transmission on occupied channels, ensuring efficient and interference-free communication. When enabled, Busy Lockout automatically monitors the channel for existing transmissions and prevents the radio from transmitting while activity is detected.

The primary purpose of Busy Lockout is to minimize interference and ensure that communications are clear and uninterrupted. By preventing users from transmitting over ongoing conversations or other radio activity, Busy Lockout helps maintain communication etiquette and prevents conflicts between users sharing the same frequency.

Busy Lockout operates by continuously monitoring the channel for incoming signals. If activity is detected, the radio temporarily disables the transmit function, preventing the user from transmitting until the channel becomes clear again. Once the ongoing transmission concludes, the radio automatically resumes normal operation, allowing users to transmit as needed.

This feature is particularly useful in environments where multiple users share the same frequencies, such as amateur radio repeaters, public safety channels, or community events. By reducing the likelihood of accidental interference, Busy Lockout helps optimize channel usage and ensures that important messages are delivered promptly and without interruption.

Advanced Features

Memory Scan for Frequently Used Frequencies

Memory scan is an advanced feature available on some Baofeng radios, designed to simplify the process of monitoring and accessing frequently used frequencies. This feature allows users to store and organize a list of preferred channels or frequencies in the radio's memory for quick and easy access.

Licensed users, such as amateur radio operators or public safety personnel, often have a need to monitor multiple frequencies simultaneously or switch between channels frequently during operations. Memory scan functionality streamlines this process by allowing users to pre-program their preferred frequencies into the radio's memory banks and scan through them at the touch of a button.

Users can configure the radio to scan through specific memory channels, frequency ranges, or bands, depending on their communication requirements. This allows for efficient monitoring of critical channels while minimizing the need for manual frequency input or adjustment.

Memory scan also enables users to prioritize certain channels or frequencies based on their importance or relevance to specific tasks or situations. By organizing frequencies into logical groups or categories within the radio's memory, users can quickly locate and access the channels they need without having to sift through unrelated frequencies.

Cross-Band Repeat: Extending Communication Range (Concept Only)

Cross-band repeat is an advanced radio operation technique that allows users to extend the communication range of their Baofeng radios by leveraging the capabilities of a repeater station. In a traditional repeater system, incoming signals are received on one frequency and re-transmitted on another frequency, typically at a higher power level or from an elevated location, to extend coverage area.

Baofeng radios equipped with cross-band repeat functionality can serve as portable repeaters, allowing users to relay signals between different frequency bands or communication networks. This feature is particularly useful in emergency situations or remote locations where traditional repeater infrastructure may be unavailable or impractical to deploy.

To utilize cross-band repeat, users must configure their Baofeng radios to operate in repeater mode and establish communication with a compatible repeater station. Once configured, the radio will automatically relay incoming signals between the selected input and output frequencies, effectively extending the communication range for users operating within the coverage area of the repeater.

While cross-band repeat offers significant benefits in terms of extending communication range and enhancing coverage, it requires careful planning and coordination to ensure compatibility with existing repeater infrastructure and compliance with regulatory requirements.

Additionally, users should be aware of potential limitations and considerations, such as power consumption, antenna selection, and frequency coordination, when deploying cross-band repeat operations.

Overall, advanced features such as memory scan and cross-band repeat expand the capabilities and versatility of Baofeng radios, providing licensed users with powerful tools for efficient and effective communication in a variety of situations. By leveraging these features, users can maximize the utility of their radios and stay connected when it matters most.

CHAPTER 5: Expanding Your Baofeng's Potential with Accessories

Enhancing Your Communication Experience: Essential Accessories

Extra Batteries for Extended Use

One of the most essential accessories for Baofeng radios is extra batteries. While Baofeng radios are known for their reliable performance and long battery life, having spare batteries on hand ensures uninterrupted communication, especially during extended outings or emergency situations.

Extra batteries provide peace of mind knowing that you'll always have a backup power source available when needed. Whether you're hiking in the wilderness, attending a day-long event, or participating in a search and rescue operation, spare batteries enable you to stay connected and maintain communication with your team without worrying about running out of power.

When selecting spare batteries for your Baofeng radio, opt for high-quality rechargeable lithium-ion (Li-ion) batteries with a capacity that matches or exceeds the original battery supplied with your radio. Consider investing in a multi-unit battery charger to conveniently charge multiple batteries simultaneously, ensuring they're ready for use when needed.

Carrying Cases for Protection and Portability

Carrying cases are essential accessories for protecting your Baofeng radio from damage and ensuring its portability during travel or outdoor activities. Designed to fit specific radio models snugly, carrying cases provide a secure and padded enclosure to shield your radio from impacts, scratches, and dust.

Whether you're hiking, camping, or working in the field, a durable carrying case keeps your Baofeng radio safe and secure, allowing you to focus on your activities without worrying about damage to your equipment. Look for cases with additional compartments or pockets to store spare batteries, antennas, and other accessories, keeping everything organized and easily accessible.

Carrying cases come in a variety of styles and materials, including nylon, leather, and hard-shell designs. Choose a case that suits your preferences and requirements, balancing protection, portability, and ease of use. Consider features such as belt loops, shoulder straps, or MOLLE compatibility for added convenience and versatility in carrying options.

Speaker/Mics or Headsets for Hands-Free Operation

Speaker/microphones (speaker/mics) and headsets are indispensable accessories for hands-free operation of Baofeng radios, allowing users to communicate effectively while keeping their hands free for other tasks. Whether you're working in a noisy environment, participating in outdoor activities, or simply seeking added convenience, speaker/mics and headsets enhance your communication experience.

Speaker/mics attach directly to the radio and feature a built-in microphone and speaker, allowing you to transmit and receive messages without having to hold the radio itself. This is particularly useful in situations where manual operation of the radio may be impractical or unsafe, such as while driving or operating machinery.

Headsets provide a more immersive communication experience, combining earpieces and a microphone boom to deliver clear audio and voice transmission directly to your ears and mouth. This helps reduce ambient noise and distractions, ensuring clear and intelligible communication even in noisy environments.

When choosing a speaker/mic or headset for your Baofeng radio, consider factors such as comfort, durability, and compatibility with your specific radio model. Look for features such as adjustable straps, reinforced cables, and noise-canceling microphones to enhance performance and usability in a variety of conditions.

By investing in essential accessories such as extra batteries, carrying cases, and speaker/mics or headsets, you can enhance your communication experience with Baofeng radios, ensuring reliability, protection, and convenience in any situation. Whether you're a casual user, outdoor enthusiast, or professional communicator, these accessories are essential tools for expanding your Baofeng's potential and maximizing its utility.

Programming Cables for Customization

Programming cables are indispensable accessories for Baofeng radios, allowing users to customize and program their radios with ease. These cables serve as a bridge between the radio and a computer, enabling users to access advanced programming features and modify radio settings using compatible software.

While software is not mentioned here, it's important to note that programming cables are typically used in conjunction with programming software specifically designed for Baofeng radios. This software provides a user-friendly interface for programming frequencies, channels, squelch settings, and other radio parameters, offering greater flexibility and control over radio operation.

Programming cables come in various configurations and connector types, depending on the specific model of Baofeng radio you own. Most Baofeng radios feature a standard Kenwood-style two-pin connector, which interfaces with the programming cable and allows for data transfer between the radio and computer.

Using programming cables and software, users can program frequencies manually, import frequency lists from online databases, and create custom channel configurations tailored to their specific communication needs. This allows for quick and efficient programming of Baofeng radios, eliminating the need for manual entry of frequency data and reducing the risk of errors.

Additionally, programming cables facilitate firmware updates and radio cloning, allowing users to transfer programming settings between multiple radios or update the radio's firmware to access new features or improvements.

Overall, programming cables are essential accessories for Baofeng radio owners who wish to unlock the full potential of their radios and customize them to suit their individual preferences and communication requirements. By enabling advanced programming capabilities, these cables empower users to take full control of their radios and maximize their utility in a variety of situations.

Optimizing Range with Antennas: Matching Your Needs

The Role of Antennas in Radio Communication

Antennas play a crucial role in radio communication, serving as the interface between the radio transmitter/receiver and the surrounding environment. They convert electrical signals into radio waves for transmission and vice versa for reception, allowing radios to communicate over varying distances and terrain types.

In essence, antennas act as the "eyes and ears" of the radio, determining its range, coverage area, and performance characteristics. The design, size, and configuration of the antenna directly impact its efficiency, radiation pattern, and ability to transmit and receive signals effectively.

Stock Antennas vs. Extended Range and Directional Antennas

Baofeng radios typically come equipped with stock antennas optimized for general-purpose communication over short to moderate distances. While these antennas are suitable for basic communication needs, they may have limitations in terms of range, coverage, and signal penetration, particularly in challenging environments or over long distances.

To optimize range and performance, many Baofeng radio users opt to upgrade to extended range or directional antennas. Extended range antennas feature a longer length and different design elements compared to stock antennas, allowing for improved signal propagation and increased coverage area.

Directional antennas, such as Yagi or beam antennas, offer even greater performance enhancements by focusing radio energy in specific directions, increasing signal strength and range in the desired direction while minimizing interference and signal loss in other directions.

Choosing the Right Antenna: Benefits, Limitations, and Safe Use

When selecting an antenna for your Baofeng radio, consider factors such as operating frequency, terrain, environment, and intended use. Choose an antenna that matches your specific communication needs and offers the desired balance of range, coverage, and performance characteristics.

Benefits of upgrading to an extended range or directional antenna include:

- Increased signal strength and range
- Improved signal clarity and reception quality
- Enhanced coverage area and penetration through obstacles
- Reduced interference and noise levels

However, it's essential to consider the limitations and practical considerations associated with antenna upgrades. Extended range antennas may be larger and less portable than stock antennas, while directional antennas require careful aiming and alignment to achieve optimal performance.

Additionally, ensure that any antenna upgrades comply with regulatory requirements and safety guidelines for radio operation. Avoid transmitting with excessively high power levels or using antennas in close proximity to sensitive electronic equipment to prevent interference and potential damage.

By choosing the right antenna for your Baofeng radio and using it responsibly, you can optimize range, coverage, and performance while enjoying reliable communication in a variety of situations. Whether you're hiking in the wilderness, participating in emergency response operations, or simply staying connected with friends and family, a quality antenna is essential for maximizing the potential of your Baofeng radio.

CHAPTER 6: Safety and Regulations: Using Your Baofeng Responsibly

Understanding Licensing Requirements: Operating Legally

Baofeng radios are powerful tools for communication, but with great power comes great responsibility. It's essential for users to operate their radios safely and in compliance with relevant regulations to ensure the integrity of the radio spectrum and the safety of themselves and others. This chapter explores important considerations for using your Baofeng radio responsibly.

Operating a Baofeng radio legally involves understanding and adhering to licensing requirements set forth by regulatory authorities. Obtaining the necessary licenses ensures compliance with regulations, promotes responsible radio operation, and helps maintain the integrity of the radio spectrum.

The Importance of Obtaining a License (e.g., FCC Amateur Radio)

In many countries, including the United States, amateur radio operators are required to obtain a license from the regulatory authority responsible for managing radio communications. In the United States, this authority is the Federal Communications Commission (FCC). Obtaining an amateur radio license from the FCC allows individuals to legally operate Baofeng radios and other amateur radio equipment on designated frequency bands.

Obtaining a license serves several important purposes:

1. Legal Compliance: Operating a Baofeng radio without a license is illegal and can result in fines, penalties, and confiscation of equipment. By obtaining a license, operators demonstrate their commitment to following regulations and operating within the bounds of the law.

2. Technical Competency: The process of obtaining a license typically involves passing an examination that tests knowledge of radio theory, operating procedures, and regulatory requirements. This ensures that licensed operators have the necessary skills and knowledge to operate their radios safely and effectively.

3. Access to Frequencies: Licensed amateur radio operators have access to a wide range of frequency bands allocated specifically for amateur use. These bands offer opportunities for experimentation, emergency communication, public service, and international communication, providing a rich and diverse radio operating experience.

Different License Types and How to Get One

In the United States, the FCC offers several different types of amateur radio licenses, each with its own privileges and requirements:

1. Technician Class: The entry-level license, which grants privileges on VHF and UHF bands.

2. General Class: Provides additional privileges on HF (shortwave) bands, allowing for long-distance communication.

3. Amateur Extra Class: The highest level of amateur radio license, offering the most extensive operating privileges across all amateur bands.

To obtain an amateur radio license in the United States, individuals must pass a written examination administered by volunteer examiners accredited by the FCC. Study materials and resources are available to help prepare for the examination, including online courses, study guides, and practice exams.

Consequences of Operating Without a License

Operating a Baofeng radio without the necessary license carries serious consequences, both legally and ethically. In addition to potential fines and penalties imposed by regulatory authorities, unlicensed operators risk causing interference to licensed users, disrupting critical communications, and undermining the integrity of the radio spectrum.

Unlicensed operation also reflects poorly on the amateur radio community as a whole, undermining the efforts of licensed operators to promote responsible radio operation, emergency preparedness, and public service. It is incumbent upon all radio users to operate within the bounds of the law and respect the licensing requirements established by regulatory authorities.

In conclusion, understanding and complying with licensing requirements is essential for operating a Baofeng radio legally and responsibly. By obtaining the necessary licenses, operators demonstrate their commitment to responsible radio operation, contribute to the integrity of the radio spectrum, and gain access to a wide range of operating privileges and opportunities within the amateur radio community.

Designated Frequencies for GMRS, FRS, MURS (Depending on Region)

In addition to emergency frequencies, other radio services such as the General Mobile Radio Service (GMRS), Family Radio Service (FRS), and Multi-Use Radio Service (MURS) have designated frequencies reserved for specific uses. Baofeng radio users should be aware of these frequency allocations and operate their radios accordingly to avoid interference with licensed users of these services.

For example, GMRS frequencies are commonly used for personal and family communication over longer distances, while FRS frequencies are limited to shorter-range communication and do not require a license. MURS frequencies are available for both personal and business use and are subject to certain restrictions on transmitter power and antenna height.

By familiarizing themselves with the frequency allocations and usage guidelines for GMRS, FRS, and MURS, Baofeng radio users can operate their radios responsibly and avoid causing interference to licensed users of these services.

Keeping Emergency Channels Clear for Critical Communication

During emergencies and disasters, clear and unobstructed communication is essential for coordinating response efforts and providing assistance to those in need. Baofeng radio users should prioritize keeping designated emergency channels clear for critical communication by refraining from unnecessary transmissions and avoiding interference with licensed users of these frequencies.

In the event of an emergency, Baofeng radio users should monitor designated emergency channels and be prepared to provide assistance or relay important information as needed. By maintaining situational awareness and cooperating with other radio operators and emergency services, Baofeng radio users can play a valuable role in supporting emergency response efforts and promoting public safety.

Battery Safety and Disposal: Protecting Yourself and the Environment

Proper handling and disposal of batteries are essential aspects of using Baofeng radios responsibly. By following safe battery handling practices and disposing of batteries properly, users can protect themselves, their equipment, and the environment from potential harm.

Safe Battery Handling Practices

Baofeng radios typically use rechargeable lithium-ion (Li-ion) batteries, which offer high energy density and long-lasting performance. However, Li-ion batteries can pose safety risks if mishandled or abused. To ensure safe operation, users should follow these battery handling practices:

- Use only manufacturer-approved batteries and chargers designed for use with Baofeng radios.

- Inspect batteries regularly for signs of damage, such as swelling, leaks, or corrosion.

- Avoid exposing batteries to extreme temperatures, moisture, or physical impact.

- Do not attempt to disassemble, puncture, or modify batteries in any way.

- Store batteries in a cool, dry place away from direct sunlight and heat sources when not in use.

By following these guidelines, Baofeng radio users can minimize the risk of battery-related accidents and ensure safe and reliable operation of their radios.

Proper Disposal Methods for Baofeng Batteries

When it comes time to dispose of old or damaged batteries, it's important to do so properly to prevent environmental contamination and comply with local regulations. Baofeng radio users should follow these disposal methods for batteries:

- Recycle batteries through authorized recycling facilities or programs that accept rechargeable batteries. Many electronics retailers, municipal waste management centers, and recycling organizations offer battery recycling services.

- Do not dispose of batteries in regular household trash or incinerate them, as this can release harmful chemicals into the environment.

- If recycling options are not available locally, contact the manufacturer or retailer for guidance on proper disposal methods.

By recycling batteries responsibly, Baofeng radio users can help reduce waste, conserve resources, and protect the environment for future generations.

In conclusion, respecting the airwaves, operating within regulatory guidelines, and practicing responsible battery handling and disposal are essential aspects of using Baofeng radios safely and responsibly. By following these principles, Baofeng radio users can contribute to a positive and cooperative radio environment while minimizing their impact on the environment and promoting public safety.

CHAPTER 7: Keeping Your Baofeng Running Smoothly: Maintenance and Troubleshooting

Basic Maintenance: Caring for Your Baofeng Radio

Maintaining your Baofeng radio is essential to ensure its longevity and optimal performance. By following basic maintenance practices, you can keep your radio in top condition for reliable communication.

Cleaning Instructions for Radio Body, Antenna, and Microphone

Regular cleaning helps prevent dirt, dust, and debris from accumulating on your Baofeng radio, which can affect its performance and appearance. Here are some cleaning instructions for different parts of your radio:

- Radio Body: Use a soft, dry cloth to wipe down the exterior of the radio body, removing any surface dust or smudges. For stubborn dirt or grime, dampen the cloth with a mild soap solution and gently wipe the surface, being careful not to let moisture seep into the internal components.

- Antenna: Clean the antenna regularly to maintain optimal signal transmission and reception. Use a soft cloth or brush to remove dirt and debris from the antenna surface. Avoid using abrasive materials or harsh chemicals that could damage the antenna coating.

- Microphone: Keep the microphone clean and free of debris to ensure clear audio transmission. Use a soft brush or compressed air to remove any dirt or dust from the microphone grille and surface. Be gentle to avoid damaging the microphone elements.

Proper Storage to Prevent Damage

Proper storage helps protect your Baofeng radio from damage and extends its lifespan. Follow these tips for storing your radio when not in use:

- Store your radio in a dry, cool environment away from direct sunlight and heat sources.

- Avoid storing your radio in excessively humid or damp conditions, as moisture can damage internal components.

- Use a protective carrying case or pouch to shield your radio from scratches, impacts, and dust when not in use.

- Remove batteries from the radio if it will be stored for an extended period to prevent corrosion or leakage.

By following these basic maintenance practices, you can keep your Baofeng radio clean, functional, and ready for use whenever you need it.

Troubleshooting Common Issues: Getting Your Radio Back on Track

Despite proper maintenance, Baofeng radios may encounter occasional issues that require troubleshooting. Knowing how to identify and address common problems can help you get your radio back on track quickly and efficiently.

Addressing Low Battery Warnings, Unclear Reception, and Button Malfunctions

Some of the most common issues Baofeng radio users may encounter include:

- Low Battery Warnings: If your radio displays a low battery warning or experiences reduced operating time, replace the batteries with fully charged ones or recharge the batteries using a compatible charger. Ensure that the batteries are properly inserted and making good contact with the battery terminals.

- Unclear Reception: If you experience unclear reception or weak signal strength, try adjusting the antenna length or orientation to improve signal reception. Move to a higher location or open area with fewer obstructions to maximize signal coverage.

- Button Malfunctions: If buttons on your Baofeng radio become unresponsive or malfunction, check for debris or moisture that may be interfering with button operation. Clean the button contacts using a soft brush or compressed air, and ensure that buttons are not stuck or damaged.

When to Seek Help from Qualified Technicians

If you are unable to resolve a problem with your Baofeng radio through basic troubleshooting, or if you suspect a more serious technical issue, it may be necessary to seek help from qualified technicians or authorized service centers. Technicians have the expertise and equipment to diagnose and repair complex issues, ensuring that your radio is restored to optimal working condition.

When seeking professional assistance, be sure to provide detailed information about the problem you are experiencing and any troubleshooting steps you have already taken. This will help technicians diagnose the issue more accurately and expedite the repair process.

In conclusion, by performing basic maintenance and knowing how to troubleshoot common issues, you can keep your Baofeng radio running smoothly and reliably for years to come. By taking care of your radio and addressing problems promptly, you can ensure that it remains a valuable tool for communication in various situations.

CONCLUSION

In this comprehensive guide, we have delved into the world of Baofeng radios, exploring their functionality, features, and the essential skills needed to operate them effectively. As we conclude our journey, let's recap the key takeaways and reflect on the power of Baofeng radios.

Recap and Key Takeaways: The Power of Baofeng Radios

Baofeng radios are powerful tools for communication, offering a wide range of features and capabilities that make them indispensable for various applications. From two-way communication in outdoor adventures to emergency preparedness and public service, Baofeng radios provide reliable and efficient communication solutions.

Throughout this guide, we have covered essential topics such as:

-Understanding Baofeng Radios: From their affordability and features to the fun they bring to radio enthusiasts, Baofeng radios offer a gateway to the world of two-way communication.

-Unveiling Your Baofeng Radio: We explored the process of unboxing, setting up, and powering up your Baofeng radio for success, emphasizing the importance of proper battery handling and basic operation.

-Demystifying Frequencies and Channels: Understanding the radio spectrum, channels, and privacy features such as CTCSS/DCS tones are crucial for optimizing communication and minimizing interference.

-*Mastering the Art of Radio Communication:* Learning radio etiquette, effective transmission techniques, and active listening skills are essential for clear and respectful communication on the airwaves.

-*Unveiling the Features of Your Baofeng Radio*: We explored essential functionalities such as dual-band/dual-display, VOX, squelch control, and advanced features like memory scan and cross-band repeat.

-*Expanding Your Baofeng's Potential with Accessories*: Enhancing your communication experience with accessories such as extra batteries, carrying cases, speaker/mics, and antennas allows for customization and optimization of range.

-*Safety and Regulations*: Operating your Baofeng radio responsibly involves understanding licensing requirements, respecting the airwaves, and practicing safe battery handling and disposal.

-Maintenance and Troubleshooting: Basic maintenance practices such as cleaning and proper storage help keep your Baofeng radio running smoothly, while troubleshooting common issues ensures uninterrupted communication.

Encouragement for Continued Learning and Responsible Use

As you embark on your journey with Baofeng radios, remember that learning is a continuous process. Continue to explore new features, expand your knowledge of radio communication, and seek opportunities for hands-on experience to hone your skills.

Responsible use of Baofeng radios is paramount to maintaining a positive and productive radio environment. Respect the airwaves, follow regulatory guidelines, and prioritize safety in all your communication endeavors. By doing so, you contribute to the integrity of the radio spectrum and promote cooperation and goodwill among radio operators.

In conclusion, Baofeng radios offer limitless possibilities for communication, exploration, and community engagement. Embrace the power of Baofeng radios, and let them be your companion in your adventures, emergencies, and everyday communication needs.

Keep Communicating, Keep Exploring, and Keep Connecting with Baofeng Radios!

GLOSSARY OF TERMS: Defining Technical Lingo

Navigating the world of Baofeng radios and two-way communication involves encountering various technical terms and jargon. To aid in understanding and comprehension, let's delve into a comprehensive glossary of terms commonly associated with Baofeng radios and radio communication.

Baofeng Radio:

Baofeng Radio: A brand of amateur radios known for their affordability, versatility, and wide range of features.

Dual-Band: Capable of operating on two separate frequency bands simultaneously, typically VHF and UHF.

Dual-Display: Equipped with two separate display screens, allowing users to monitor multiple channels or frequencies simultaneously.

Radio Spectrum

Radio Spectrum: The range of frequencies used for radio communication, divided into various bands for different applications.

VHF (Very High Frequency): Radio frequencies ranging from 30 MHz to 300 MHz, commonly used for line-of-sight communication over relatively short distances.

UHF (Ultra High Frequency): Radio frequencies ranging from 300 MHz to 3 GHz, suitable for both line-of-sight and non-line-of-sight communication over shorter distances compared to VHF.

Frequency: The number of oscillations per second of a radio wave, measured in Hertz (Hz).

Wavelength: The distance between successive peaks or troughs of a radio wave, inversely proportional to frequency.

Channels and Frequencies:

Channel: A pre-programmed frequency or set of frequencies assigned for specific uses, such as simplex communication, repeater operation, or emergency channels.

Frequency Allocation: The process of assigning specific frequency ranges to different radio services, regulated by government agencies to prevent interference and ensure efficient use of the radio spectrum.

CTCSS (Continuous Tone-Coded Squelch System): A sub-audible tone transmitted along with the voice signal to selectively mute or open a receiver's audio, reducing interference from other users on the same frequency.

DCS (Digital-Coded Squelch): A digital encoding scheme used to selectively mute or open a receiver's audio based on a predefined digital code, providing more secure and reliable squelch control compared to CTCSS.

Communication Techniques:

Push-to-Talk (PTT): A button or switch used to activate the transmitter for transmitting voice or data signals.

Voice Operated Transmit (VOX): A feature that allows the radio to automatically transmit when it detects voice input through the microphone, enabling hands-free operation.

Squelch Control: A circuit that mutes the audio output of a receiver when no signal is present, reducing background noise and interference.

Memory Scan: Automatically scans and monitors pre-programmed frequencies or channels stored in the radio's memory.

Cross-Band Repeat: A feature that allows a radio to receive a signal on one frequency band and simultaneously retransmit it on another frequency band, extending communication range.

Safety and Regulations:

FCC (Federal Communications Commission): The regulatory agency responsible for overseeing radio communication in the United States and enforcing regulations related to spectrum allocation, licensing, and equipment standards.

Licensing Requirements: Legal mandates for obtaining a license or authorization to operate radio equipment, typically administered by government agencies to ensure competency and compliance with regulations.

Emergency Channels: Designated frequencies or channels reserved for emergency communication by public safety agencies, emergency responders, and amateur radio operators during emergencies and disasters.

Battery Safety: Practices and guidelines for handling, charging, and disposing of batteries safely to prevent accidents and environmental contamination.

Maintenance and Troubleshooting

Maintenance: Routine tasks performed to keep radio equipment in optimal condition, including cleaning, inspection, and proper storage.

Troubleshooting: The process of identifying, diagnosing, and resolving problems or malfunctions in radio equipment, often involving systematic testing and analysis of components and systems.

Low Battery Warning: An alert indicating that the battery voltage has dropped below a certain threshold, typically displayed on the radio's screen to prompt the user to recharge or replace the batteries.

Button Malfunctions: Issues with the operation or responsiveness of buttons on the radio, often caused by dirt, debris, or mechanical wear.

This glossary serves as a reference guide to help users navigate the technical terminology and concepts associated with

Baofeng radios and radio communication. By understanding these terms, users can enhance their knowledge, troubleshoot issues effectively, and communicate more efficiently in various scenarios. As technology evolves and new features emerge, staying informed and familiar with technical lingo remains essential for radio enthusiasts and operators alike.